BUG BOOK
for Kids

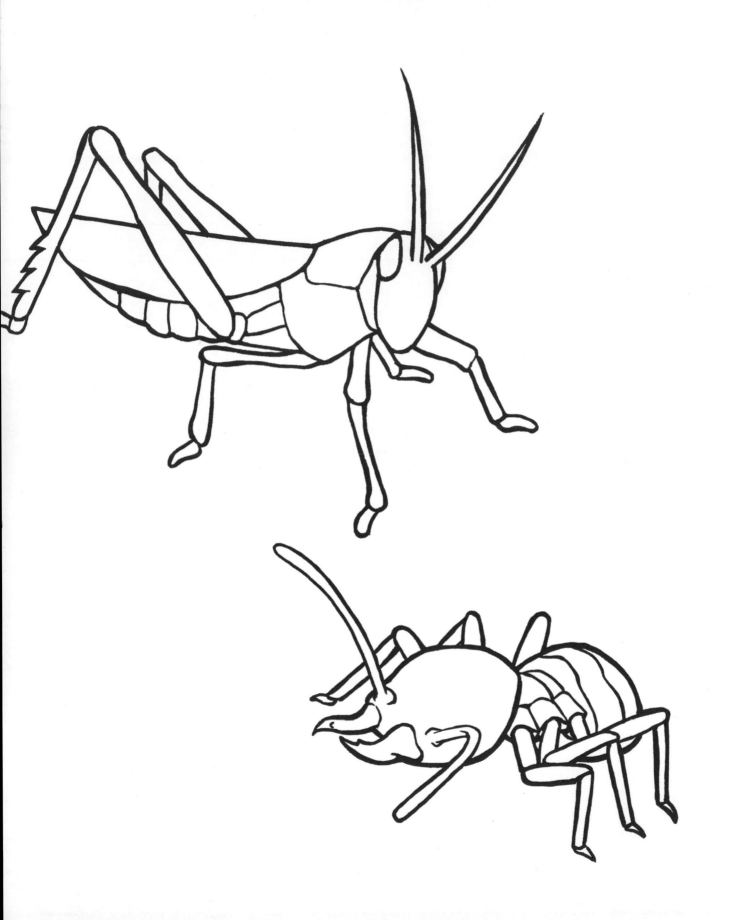

Bug Book
for Kids

Coloring Fun and Awesome Facts

BY KATIE HENRIES-MEISNER

ART BY ANDRE SIBAYAN

Z KIDS · NEW YORK

Published in the United States by Z Kids, an imprint of Zeitgeist™, a division of Penguin Random House LLC, New York.

zeitgeistpublishing.com

Zeitgeist™ is a trademark of Penguin Random House LLC

ISBN: 9780593196861

Illustrations by Andre Sibayan

Book design by Aimee Fleck

Manufactured in China

5th Printing

To DC, AR,
and of course,
Harrison.

Meet the Backyard Bugs!

Do you like bugs? Do you like to watch them crawl and jump and fly? Then you'll love coloring these awesome bugs that you might find in your own backyard!

All 25 of our backyard bugs are arthropods. The scientists who study them, called entomologists, sort arthropods into groups such as insects, arachnids, and crustaceans—all creepy-crawly animals that are fun to learn about and color.

Mostly insects fill these coloring pages. Insects have three parts to their bodies—head, thorax, and abdomen. They also usually have two pairs of wings, six legs, and a pair of antennae on their heads. True bugs are a kind of insect with a particular mouth part and unique wings. In this book, cicadas, stink bugs, and water striders are true bugs.

Many backyard bugs have life cycles that involve a complete change, or metamorphosis. They begin as eggs that hatch into larvae, which look like worms, then move into a resting stage called a pupa, where they

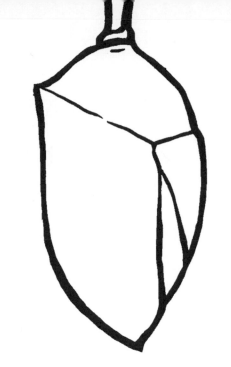

develop their legs and wings. Finally, they emerge as a full-grown adult. You may have seen this change when a caterpillar's pupa (called a chrysalis) brings forth a beautiful butterfly.

With over a million different species, backyard bugs come in all shapes, sizes, and colors. Have fun learning amazing facts about the bugs and coloring them any way you want. Before you're done, head outside with your trading cards from the back of the book to try to find and identify the bugs you've learned about and colored here. What colors and details do you notice?

And remember, new bugs are discovered

all the time, so maybe your creative coloring will match a crawling critter that's yet to be discovered!

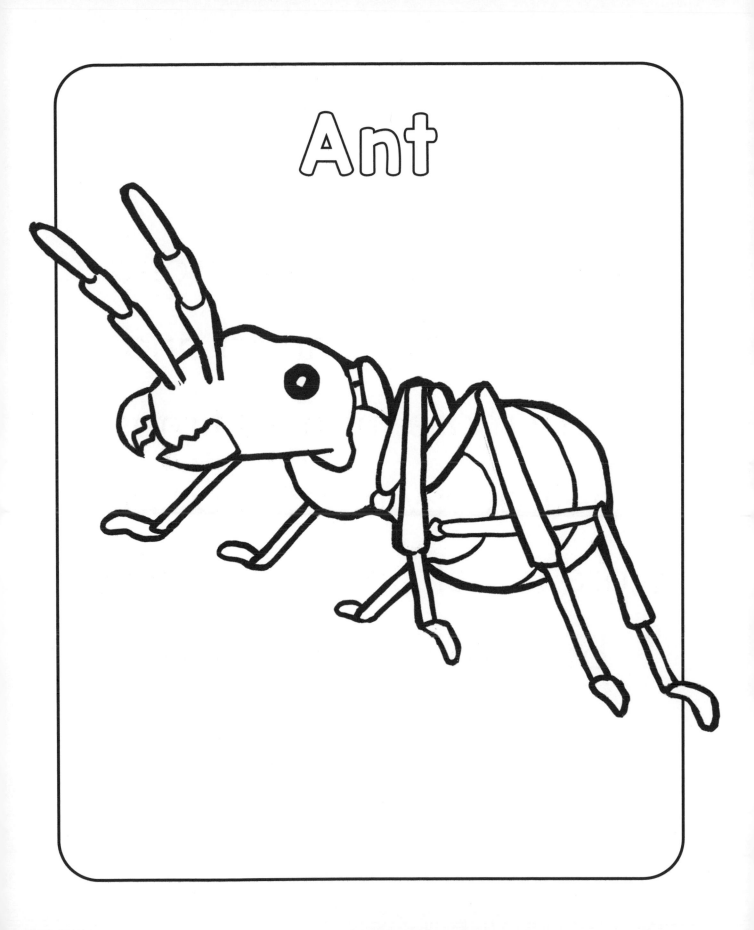

Ants are extremely strong insects. They can lift and carry more than 20 times their own weight. If a kid were this strong, they would be able to carry a car!

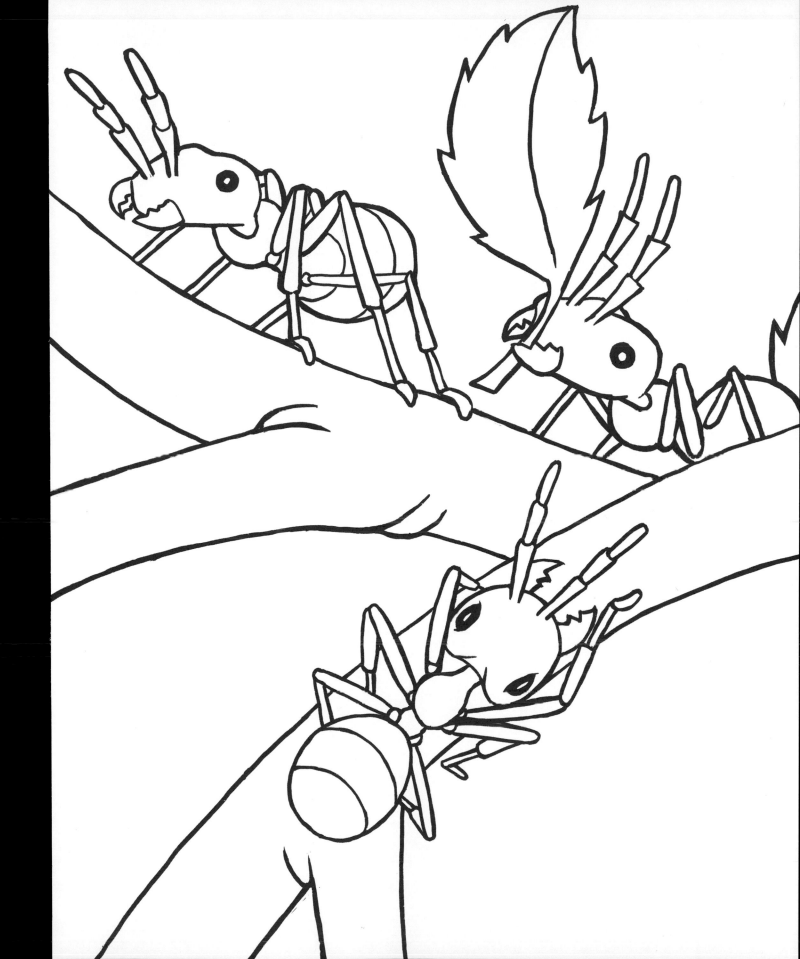

Bee

These buzzing insects in a hive include the **queen bee** and her royal subjects: female **worker bees** and male **drones**. Though they all share the hive in the summer, the female workers kick the nonworking male drones out into the cold as winter approaches.

Beetle

Beetles are the largest insect order, or group. There are over 350,000 different types of beetles. They have been on Earth for at least 230 million years, meaning they were here even before dinosaurs.

Butterfly

A caterpillar changes into a **butterfly** through a process many insects experience called metamorphosis. It takes at least 10 to 15 days for a caterpillar to change into a butterfly inside its chrysalis.

Centipede

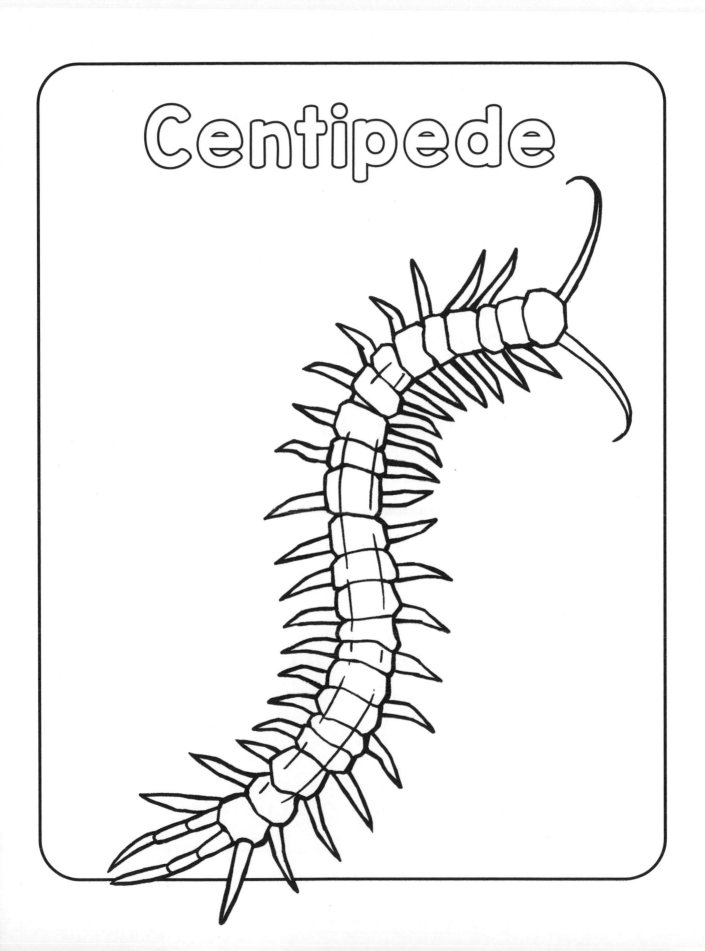

Although **centipede** means "100 legs," centipedes never have exactly 100 legs. They can have anywhere from 15 to 177 pairs of legs, depending on the kind. Centipedes can lose some of their legs to predators, but they are often able to regenerate, or regrow, new legs!

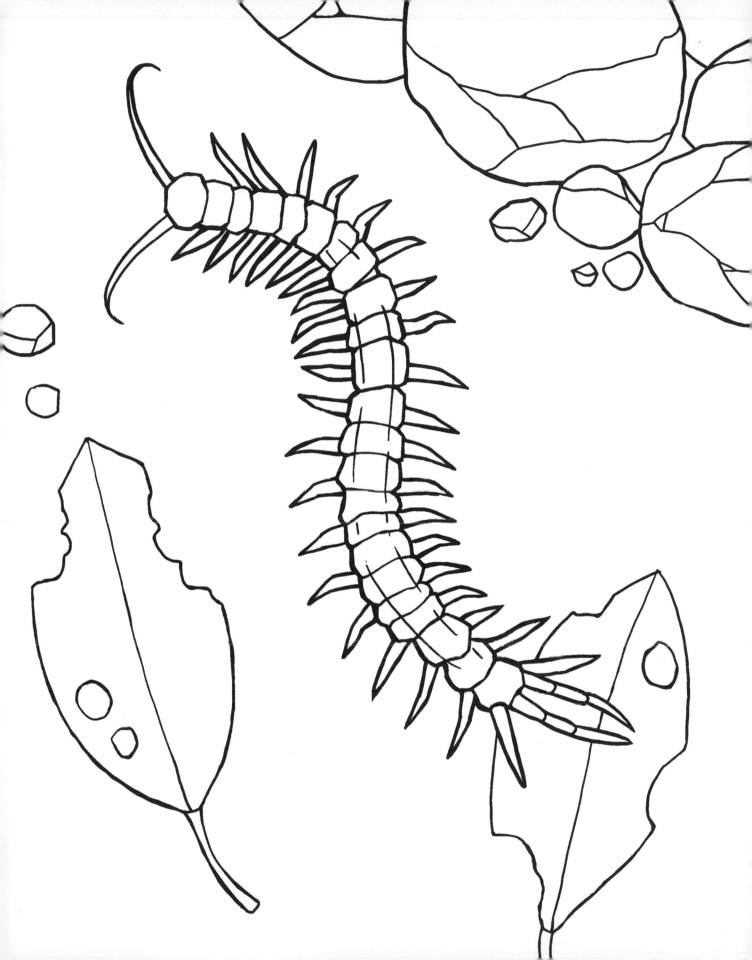

Cicada

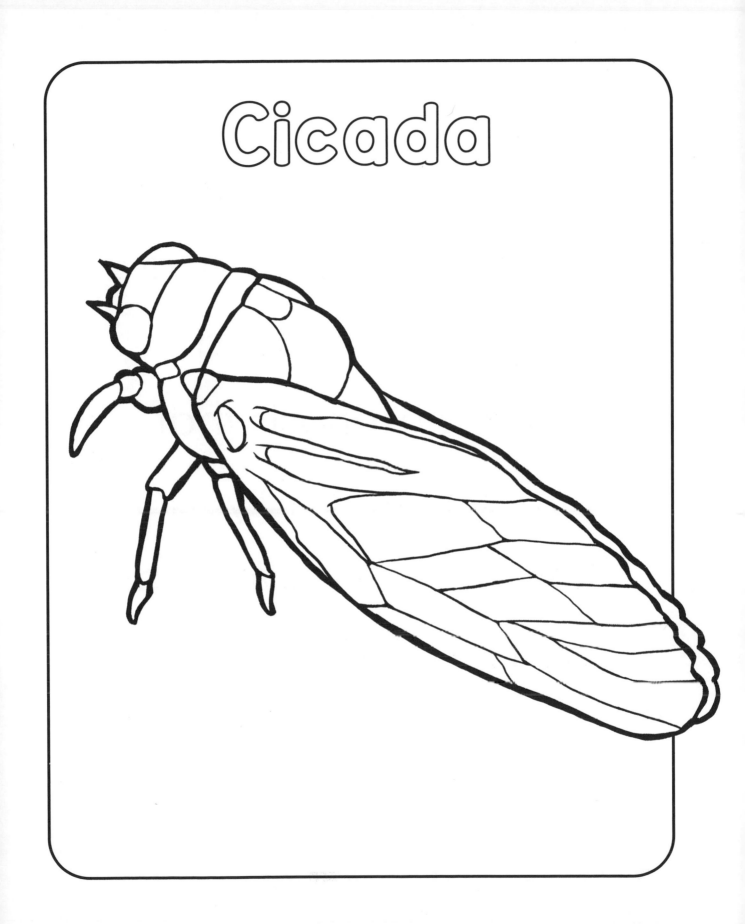

Cicadas have lengthy life cycles that involve an underground metamorphosis that can take up to 10 years. Their entire life cycle can be 13 to 17 years long. That's an old bug!

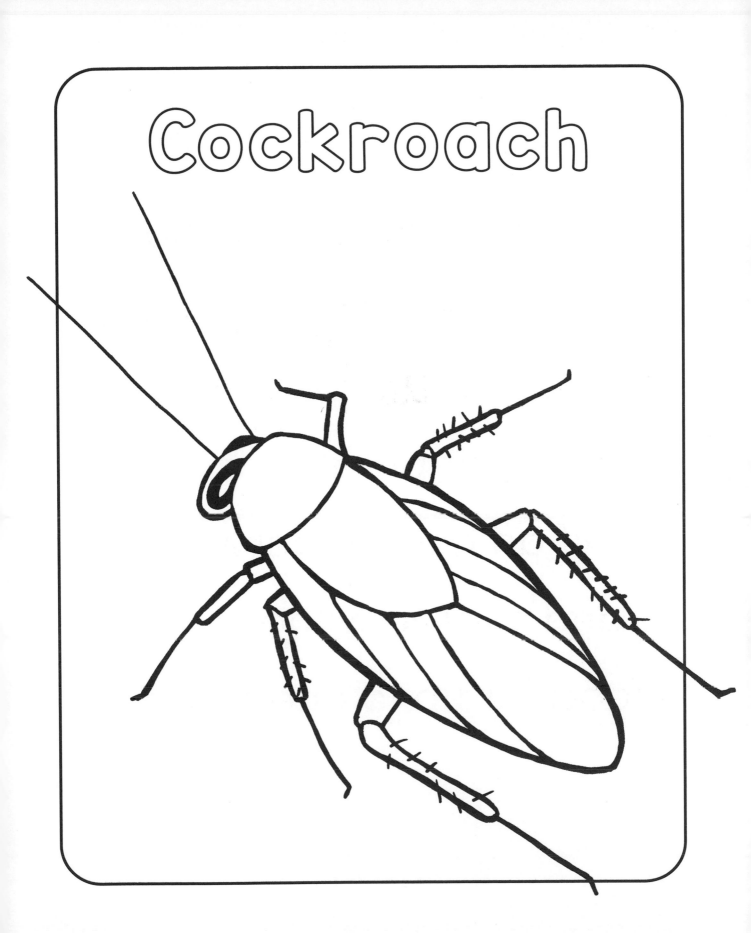

Cockroaches are survivors. These insects can withstand freezing temperatures, hold their breath for over half an hour underwater, and survive without any food for up to one month.

Cricket

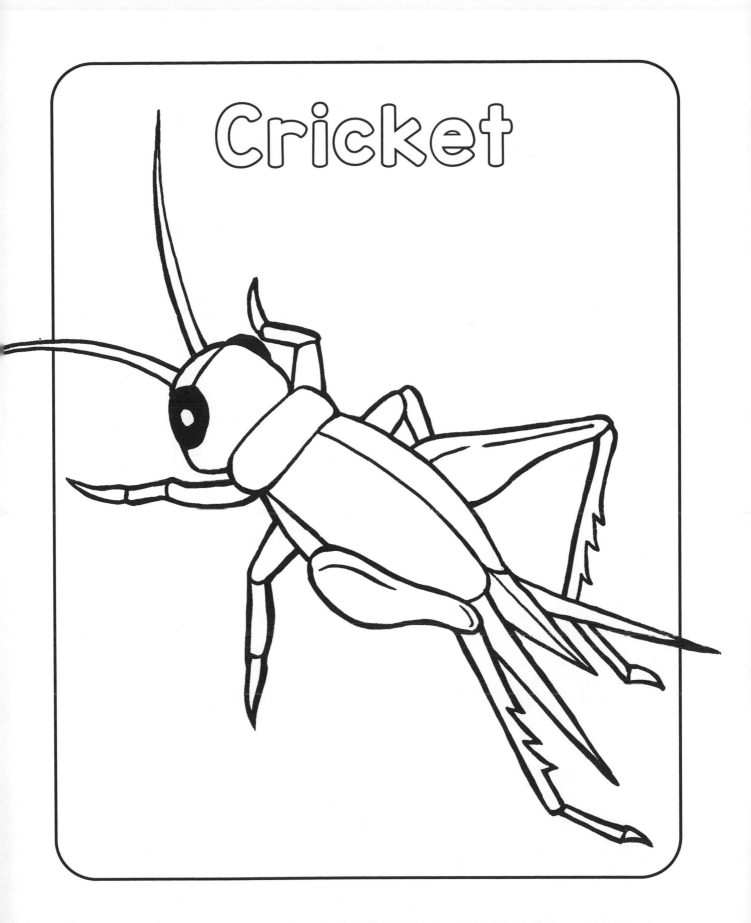

Crickets are insects that make their sound by rubbing their wings together. Their wings are covered in grooves called teeth. When you run your finger along the teeth of a comb, you make the same kind of sound.

Dragonfly

Dragonflies have better eyesight than humans, which makes them excellent hunters. Once these insects catch their prey, they eat it as they fly through the sky.

Flea

Fleas don't fly, but they are incredible jumpers, which helps them get around. They can jump well over 100 times their own length. If you could jump like these insects, you'd be able to leap over the Statue of Liberty!

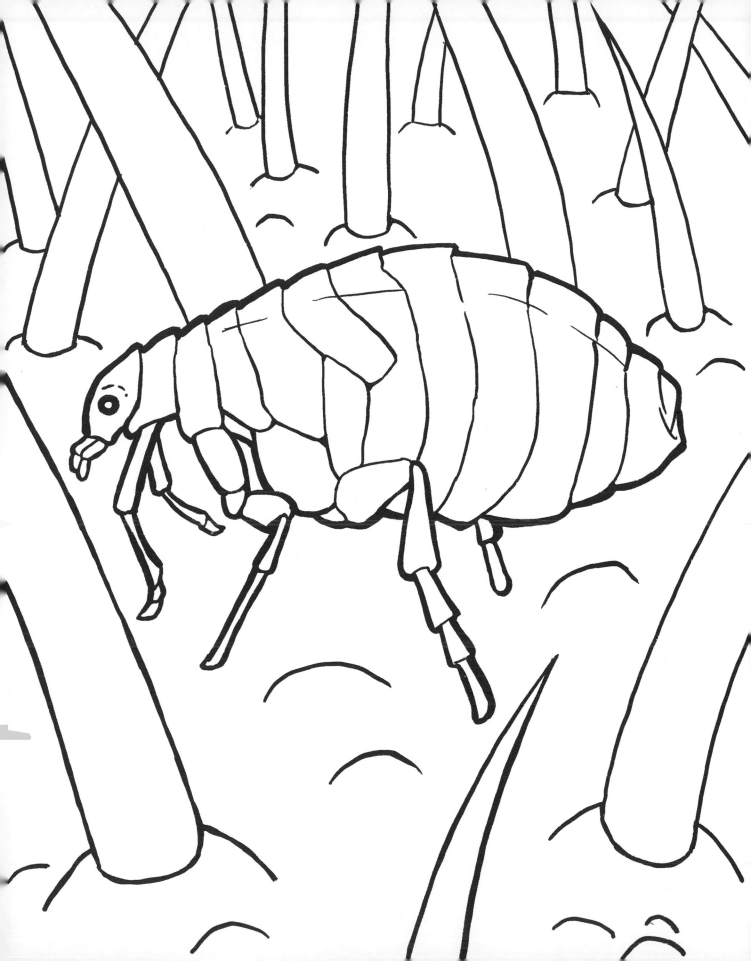

Grasshopper

Grasshoppers are often green or brown in color, but some species of this insect are very colorful. There is even one brightly colored kind called a painted grasshopper.

Housefly

Houseflies are insects that can walk upside down, taste with their feet, and see things that are behind them. A female housefly can lay up to 600 eggs in her short lifetime. That's a lot of baby flies!

Ladybug

Despite their name, **ladybugs** can be male or female. They may have up to 20 spots, stripes, or no markings at all. Their color tells predators to stay away—that insect won't taste good.

Lightning Bug

Lightning bugs (also called fireflies) are named for the light they produce, which can be green, yellow, or orange. Lightning bug larvae sometimes glow underwater or underground. This tells predators that these insects do not taste good, and it helps keep them safe.

Mosquito

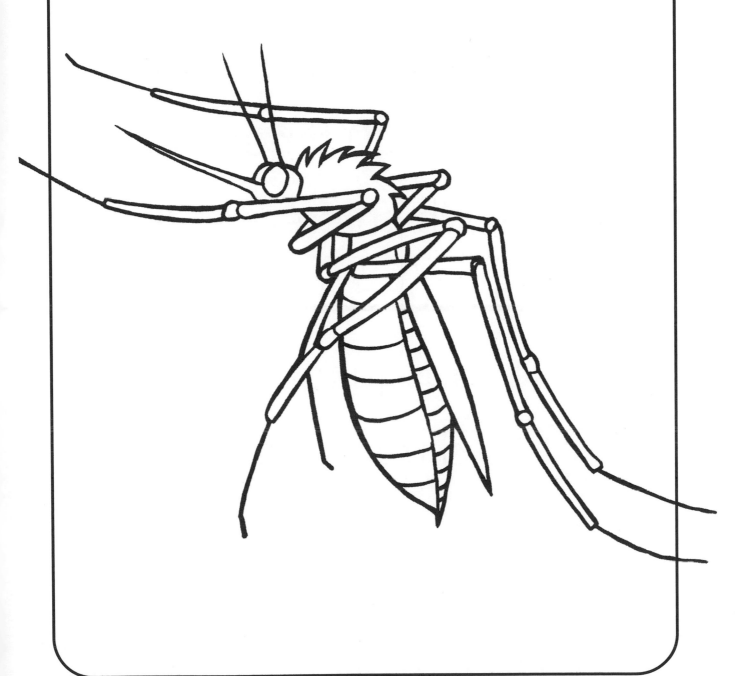

Mosquitoes are insects that can beat their tiny wings nearly 1,000 times per second, which creates their buzzing sound. Female and male mosquitoes make slightly different buzzing tones with their wings.

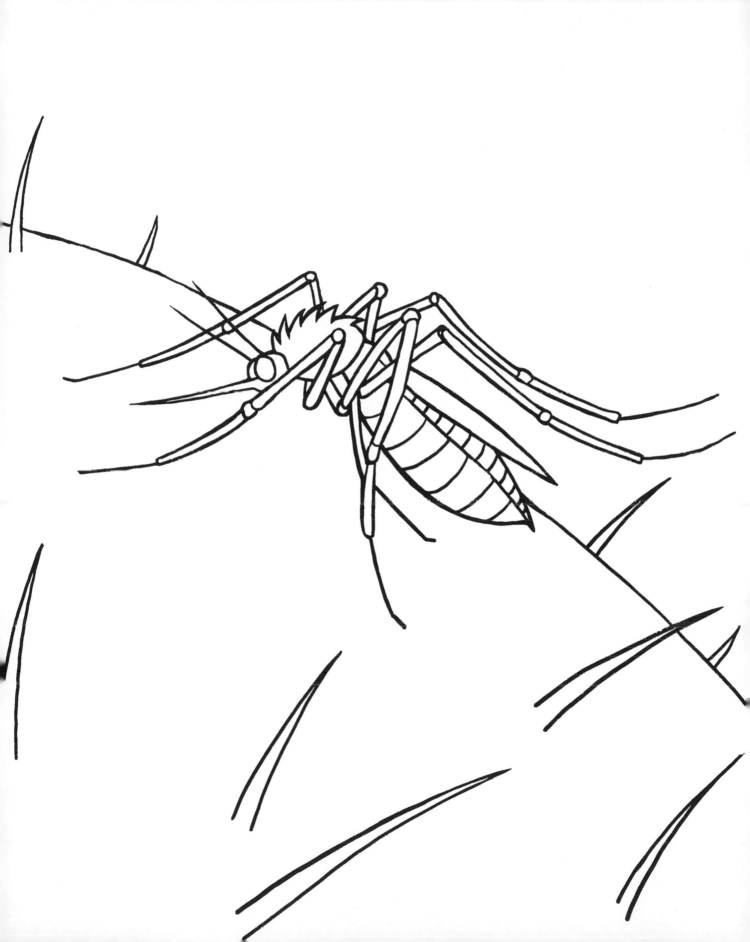

Pillbug

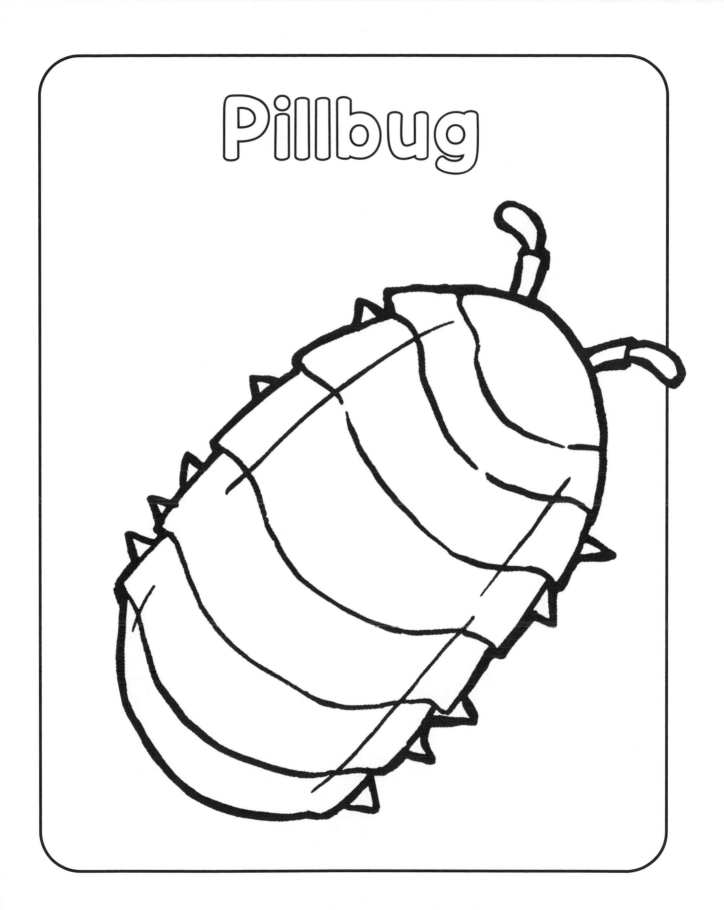

Pillbugs, or roly polies, are not actually bugs or insects at all—they are crustaceans, part of the same family as lobsters, crabs, and shrimp! Pillbugs breathe through gills, not lungs, which is why they must live in a damp place to survive.

Praying Mantis

The **praying mantis** is named for the way its two bent forearms are shaped: when they come together, this insect looks like it is praying. It has only one ear, which is located on the underside of its body.

Scorpion

Scorpions are arachnids, like spiders, though they look more like small lobsters. They have venomous stingers on their curved tails that they use to paralyze, or freeze, their prey. They also use their tail as defense in an attack.

Spider

Spiders are not bugs or insects; they are arachnids. They have two main body parts and eight legs. They come in all shapes and sizes—from the foot-long Goliath spider to the tiny Patu marplesi spider, which is smaller than the period at the end of this sentence.

Stick Insect

Stick insects are sometimes called walking sticks. They can be hard to find because they look exactly like twigs . . . until they walk away, that is! Even their eggs blend into nature; they look like seeds and are spread on the forest floor in the same way seeds are.

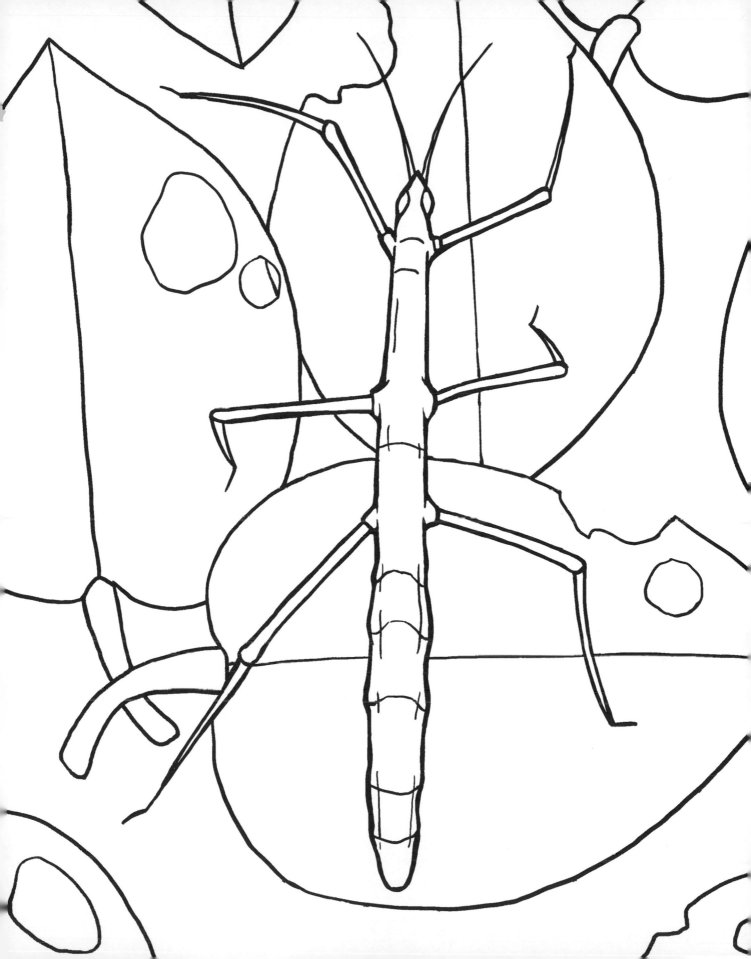

Stink Bug

Stink bugs are the skunks of the insect world. A gland in their legs gives off a nasty-tasting and nasty-smelling chemical that scares away anything that might want to eat it.

Termite

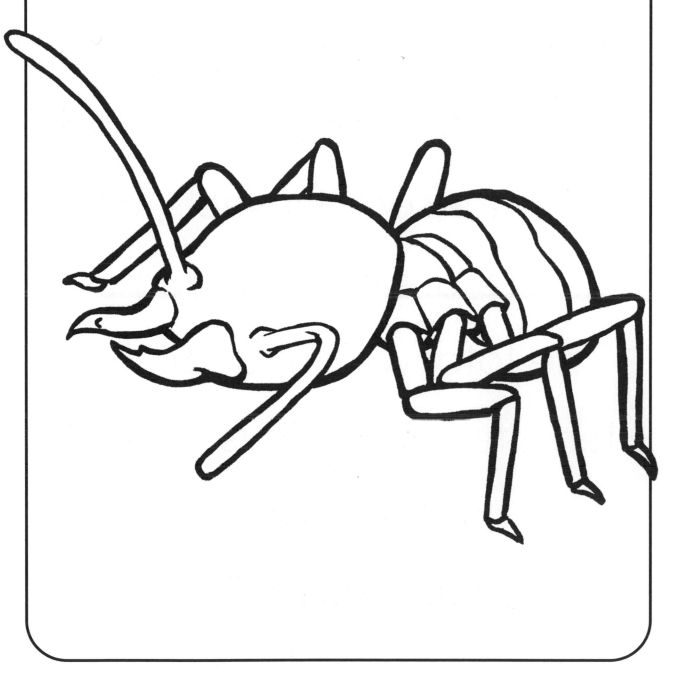

Termites are known as "silent destroyers" because they can quickly chew through wood without being noticed. They are social insects, and there are a lot of them. The total weight of all the termites on Earth is MORE than the total weight of all the humans on Earth!

Tick

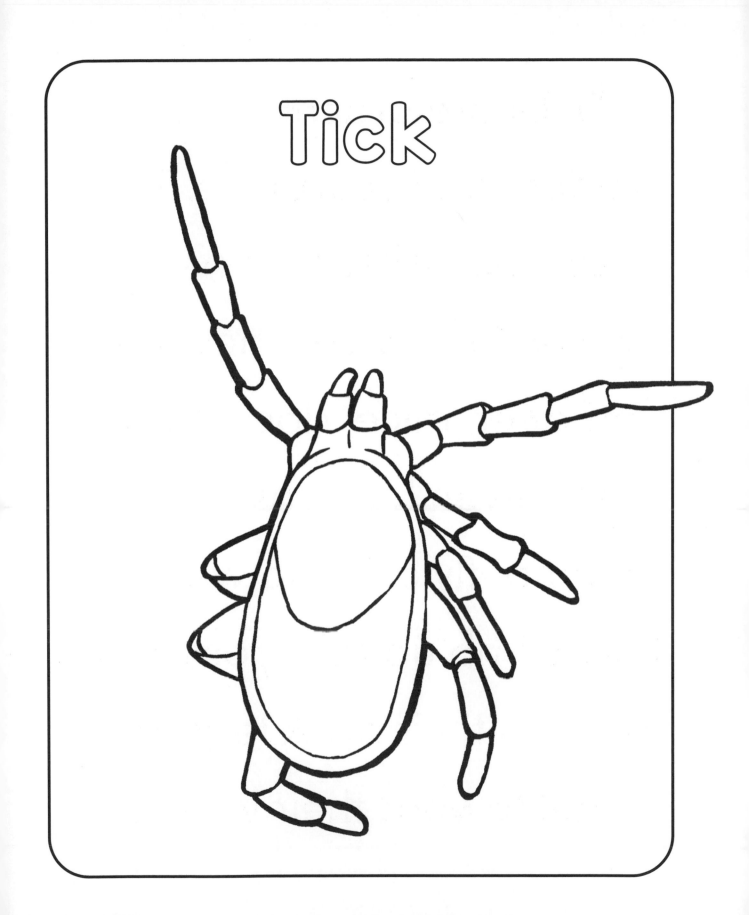

Ticks are arachnids, like spiders and scorpions. They are parasites that need blood to survive. They feed off deer, mice, dogs, birds, and even humans. Many cannot fly or jump well, so they climb up tall grass, hold on with their back legs, and fall toward their victim.

Water Strider

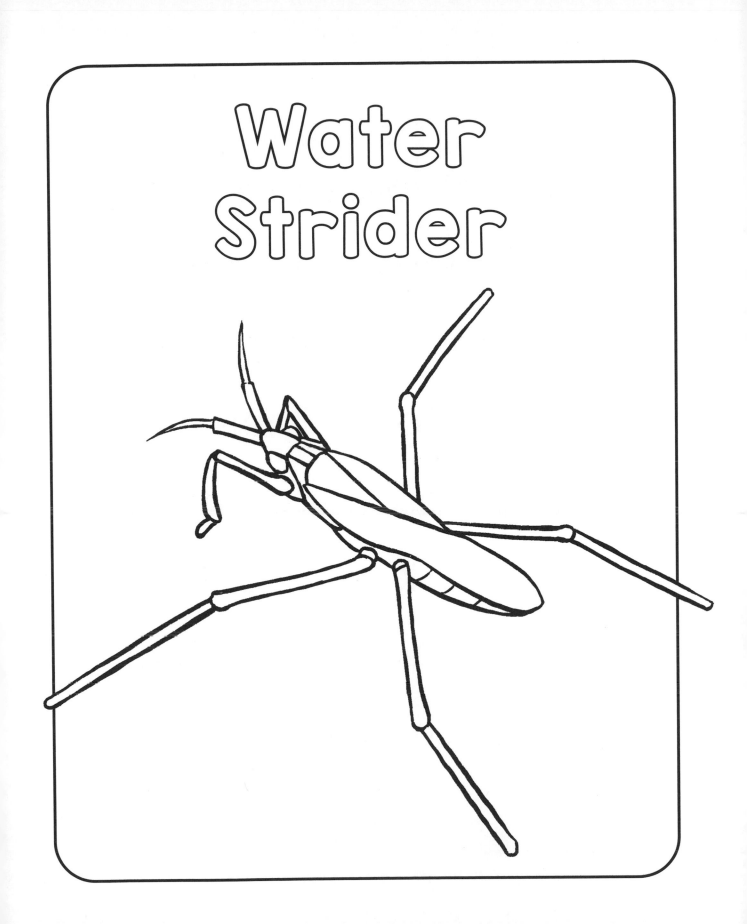

Water striders are bugs with waterproof hairs on their legs that help them walk on water. Their front legs grab prey, their middle legs work as paddles, and their back legs steer, balance, and brake.

Weevil

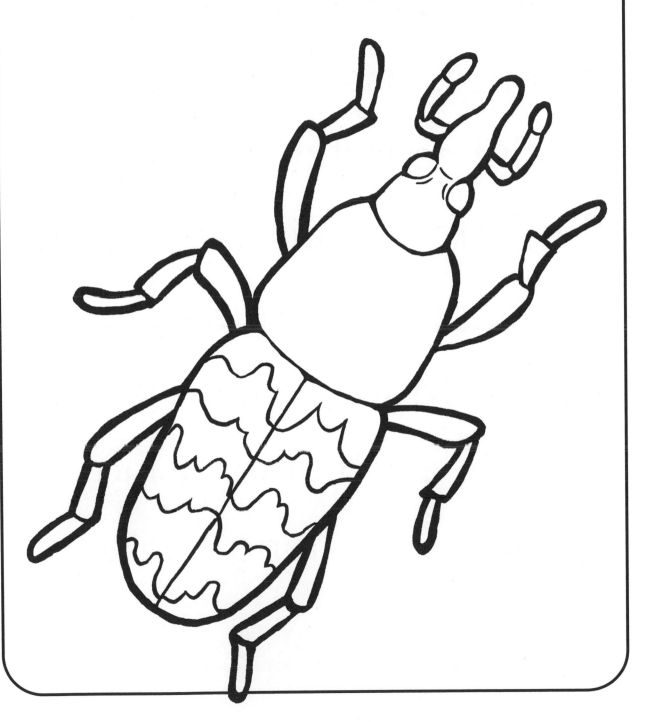

Weevils are plant-eating beetles, insects with a long snout called a rostrum on the front of their head. They make up the largest beetle family. Male giraffe weevils fight each other using their extra-long necks.

About the Author

Katie Henries-Meisner is a teacher, school leader, mom of two kids, and the author of *Dinosaur Book for Kids: Coloring Fun and Awesome Facts*. She's taught first, third, fourth, fifth, and sixth grades; fourth grade is her favorite! She grew up in suburban Massachusetts, where she developed a passion for urban education and social justice, along with a love of learning through nature, exploration, and projects. She currently lives in Northern California with her family.

About the Artist

Andre Sibayan is an illustrator and spicy food enthusiast. He created the illustrations for *Dinosaur Book for Kids: Coloring Fun and Awesome Facts*. When he's not drawing, designing, or tasting new hot sauces, he enjoys exploring nature and all of its curious creatures with his son, Niko.

Ant

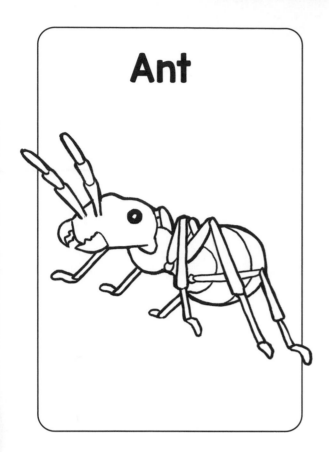

Bee

Beetle

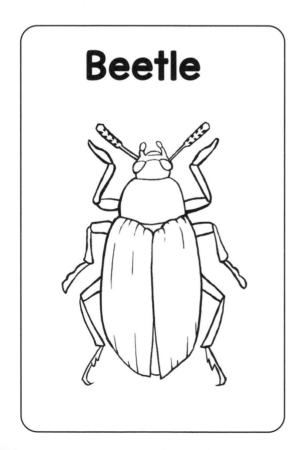

Butterfly

Centipede

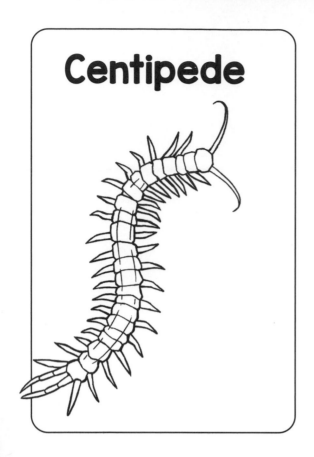

Cicada

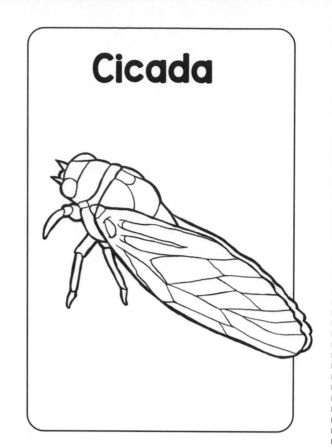

Cockroach

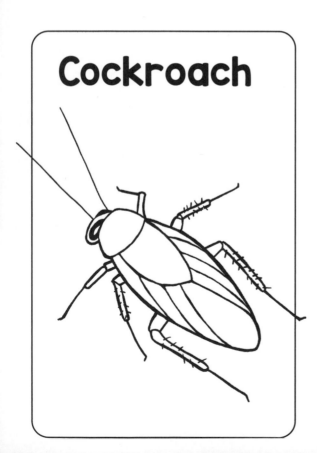

Cricket

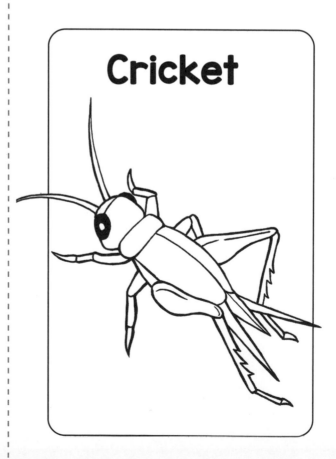

Dragonfly

Flea

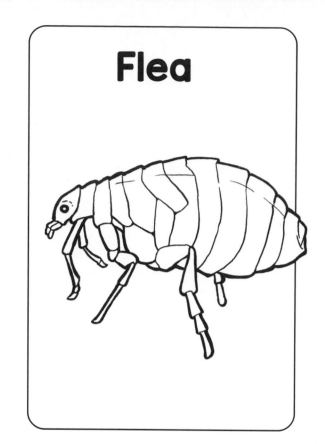

Grasshopper

Housefly

Ladybug

Lightning Bug

Mosquito

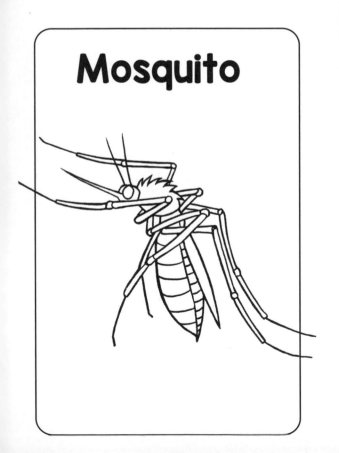

Pillbug

Praying Mantis

Scorpion

Spider

Stick Insect

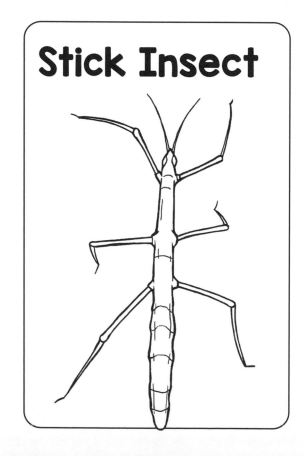

Stink Bug

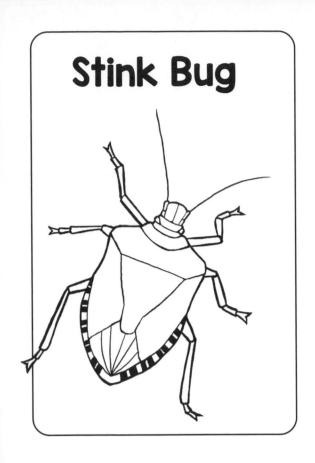

Termite

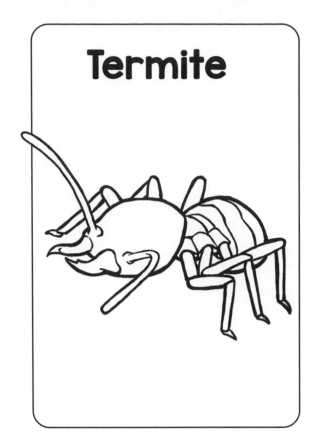

Tick

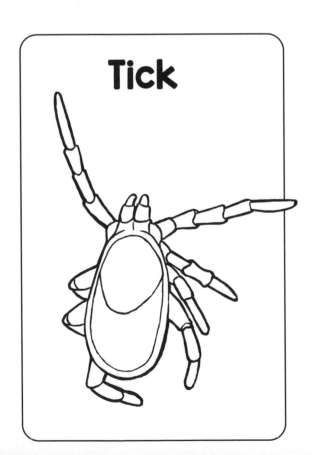

Water Strider

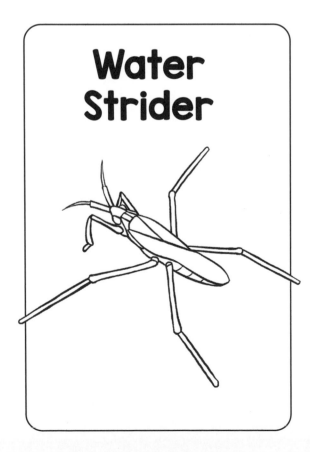

Weevil